HOW TO USE A METAL DETECTOR

Uncovering Hidden Treasures: A Beginner's Guide to Metal Detecting

Fitzpatrick J. Thompkins

Copyright © 2024 by **Fitzpatrick J. Thompkins**

All rights reserved

No part of this publication may be reproduced, stored in a retrieval system, or transmitted, in any form or by any means, electronic, mechanical, photocopying, recording, or otherwise, without the prior written permission of the author.

The information in this ebook is true and complete to the best of our knowledge. All recommendation are made without guarantee on the part of author or publisher. The author and publisher disclaim any liability in connection with the use of this information.

Table of contents

Table of contents	3
Introduction	5
Purpose of Using a Metal Detector	7
Brief History of Metal Detectors	8
Types of Metal Detectors	9
Chapter: 1 Understanding Metal Detector Technology	10
Basic Operating Principles	10
Types of Detection Technology	12
Key Components of a Metal Detector	14
Chapter: 2 Setting Up Your Metal Detector	15
Assembly Instructions	15
Initial Calibration and Settings	17
Understanding and Adjusting Sensitivity, Discrimination, and Ground Balance	19
Chapter: 3 Techniques for Successful Metal Detecting	21
Searching Techniques	21
Target Identification	23
Digging and Retrieving Items	25
Metal Detecting Etiquette	27
Chapter: 4 Choosing the Right Equipment	29
Selecting the Best Metal Detector for Your Needs	29
Essential Accessories for Metal Detecting	31
Chapter: 5 Maintenance and Care	32

Routine Maintenance Tips	32
Troubleshooting Common Problems	34
Long-term Storage Solutions	36
Chapter: 6 Legal and Ethical Considerations	**37**
Understanding Local Laws and Regulations	37
Respecting Private and Public Properties	39
Promoting Responsible Metal Detecting	40
Chapter: 7 Advanced Techniques and Tips	**42**
Enhancing Detection Depth and Accuracy	42
Using Metal Detectors in Different Environments	44
Participating in Competitions and Events	46
Chapter: 8 Applications of Metal Detecting	**48**
Treasure Hunting	48
Historical Research	50
Geological Exploration	52
Recovery and Rescue Missions	54
Chapter: 9 Case Studies and Success Stories	**56**
Notable Finds	56
Interviews with Experienced Metal Detecting Enthusiasts	57
Chapter: 10 Resources and Further Reading	**59**
Recommended Books and Guides	59
Useful Websites and Forums	61
Training and Certification Programs	63
Conclusion	**65**

Introduction

As the sun began to dip below the horizon, casting a golden hue across the shoreline, Tom Parker stepped onto the beach with his brand-new metal detector. He was an amateur treasure hunter, and today marked the beginning of his metal-detecting journey. His enthusiasm was palpable, but as he swept the detector back and forth across the sand, he quickly realized that enthusiasm alone wouldn't be enough.

An hour passed with nothing to show for his efforts—just a couple of bottle caps and a rusty nail. Frustration crept in. That's when he noticed a seasoned detectorist nearby, methodically uncovering item after item: coins, jewelry, and even an antique watch. Tom approached him, hoping for some advice.

The man, named Eddie, was a veteran in the field with over twenty years of experience. With a kind smile, he pointed to Tom's metal detector and asked, "Did you just start with this hobby?" After Tom nodded, Eddie said, "I see you're struggling a bit. There's actually a book I think you might find really helpful. It's called 'How to Use a Metal Detector.' It taught me everything when I was starting out."

Intrigued, Tom decided to give the book a shot. He purchased it the very next day, and as he flipped through the pages, he realized what he had been missing. The book was more than just a manual; it was a deep dive into the art and science of metal detecting. Each chapter was meticulously crafted to guide a novice from the simplest steps to more advanced techniques.

The first chapter explained the different types of metal detectors and their specific uses, helping Tom understand why his device was ideal for beaches but might not be the best for mineralized soils. The technical descriptions were surprisingly clear, with diagrams showing how to adjust settings like sensitivity and discrimination, crucial for ignoring junk metal.

Eager to test his new knowledge, Tom returned to the beach. This time, he adjusted his metal detector's settings according to the book's recommendations, paying close attention to the signals' nuances. The difference was night and day. He found a silver ring and several old coins, much to his delight.

The chapter on 'Searching Techniques' was a game changer. It taught him search patterns, how to maximize ground coverage, and the importance of slow and methodical sweeps. Tom also appreciated the sections on ethical metal detecting, which included tips on obtaining permissions, respecting property, and promoting responsible practices.

As the weeks turned into months, Tom's skills grew. He even started participating in forums and community digs. The book had chapters on these as well, including stories of successful treasure hunters, which kept his enthusiasm burning.

One evening, as Tom unearthed a particularly rare coin from the 1800s, a beginner approached him, detector in hand, looking every bit as lost as Tom had felt on his first day. Smiling, Tom recommended "How to Use a Metal Detector." He explained, "This isn't just a guide, it's your companion in this adventure. It'll teach you not only how to use a metal detector but how to love this hobby."

As the new enthusiast thanked him and walked away, Tom couldn't help but feel grateful for having found that invaluable book. It hadn't just made him a better treasure hunter—it had opened up a world of history, discovery, and community that he would have missed out on otherwise. And as the sun set, Tom felt connected to a past that only he, through his newfound skills and trusty metal detector, could unearth and bring back to life.

Purpose of Using a Metal Detector

Metal detectors are fascinating devices that serve a range of purposes, each adding a unique dimension to the practice of metal detecting. These versatile tools are most commonly associated with treasure hunting, which is often considered the heart of metal detecting. Enthusiasts and professionals alike explore beaches, old battlefields, abandoned homesteads, and other historical sites, aiming to uncover hidden artifacts, coins, jewelry, and sometimes even gold. The thrill of discovering something valuable or historically significant that has been buried for decades or centuries is a powerful motivator.

Beyond treasure hunting, metal detectors are also invaluable in other practical applications. For example, they play a critical role in landmine removal and unexploded ordnance disposal. In post-conflict regions around the world, metal detectors help to clear lands that are contaminated by remnants of war, thereby making them safe for civilians. This not only saves lives but also enables displaced people to return to their homes and use their land for agriculture or development.

In the field of construction and renovation, metal detectors are used to locate hidden pipes and wires within walls and underground. This helps to prevent accidental damage during construction activities, which can lead to costly repairs and dangerous situations, such as gas leaks or electrical shorts.

Archaeologists also rely heavily on metal detectors to locate and map out potential sites of interest. By detecting metal artifacts, they can establish the perimeters of ancient settlements or battlegrounds without extensive excavation, preserving the site's integrity while still uncovering important historical data.

Furthermore, metal detectors are used by law enforcement and security personnel to ensure public safety. Airports, concert venues, and public events often require people to pass through metal detectors to prevent the carrying of weapons or other illegal items. This security measure has become a routine part of safety protocols worldwide.

Finally, metal detecting can be a relaxing hobby that connects people with the outdoors. It encourages physical activity, provides educational opportunities, and can be a social activity, bringing together communities of enthusiasts who share finds, tips, and techniques.

In conclusion, the use of metal detectors spans from leisure and historical discovery to practical applications in safety and construction. Each of these purposes not only justifies the need for effective metal detecting skills but also highlights the diverse appeal of this engaging activity. Whether driven by curiosity, a love of history, or practical necessity, learning to use a metal detector effectively opens up a world of possibilities and discoveries.

Brief History of Metal Detectors

The history of metal detectors is a fascinating journey that intertwines with the evolution of technology and human curiosity. The invention of the metal detector is often credited to Alexander Graham Bell in 1881. Bell created a device to locate a bullet lodged in the chest of President James Garfield. Although his attempt was unsuccessful due to the metal bed frame interfering with the device, this marked the beginning of the use of electromagnetic fields for metal detection.

During the early 1900s, many scientists and inventors experimented with metal detection technologies, but it wasn't until the 1920s and 1930s that metal detectors became more practical and reliable. Gerhard Fischer, an electrical engineer, filed the first patent for a portable metal detector in 1925. Fischer's design, which was the first to use radio frequency to differentiate between different types of metals, laid the foundation for modern metal detecting equipment.

The use of metal detectors expanded during World War II. They were used extensively to find land mines, unexploded bombs, and other metallic objects buried in the ground. This military application helped refine the technology, making it more sensitive and reliable. Post-war, the technology was adapted for other uses, including mineral prospecting, archaeological digs, and security screening.

In the 1960s and 1970s, as electronic components became smaller and cheaper, metal detectors became more accessible to the general public. This led to a surge in popularity among hobbyists and treasure hunters. The development of the Very Low Frequency (VLF) technology during this time revolutionized the hobby. VLF detectors could discriminate between different types of metals, allowing users to ignore junk and focus on more valuable finds. This technology is a critical aspect of metal detecting that enthusiasts must master to use their equipment effectively.

Today, metal detectors are sophisticated instruments that use advanced technologies like Pulse Induction and multi-frequency detection to locate a wide array of metallic objects from relics buried in deep soil to coins on busy beaches. Learning about the historical development of these devices provides users not only with a better appreciation of their hobby but also insights into how best to utilize their equipment. For example, understanding the specific capabilities and limitations of different detector technologies can greatly enhance an enthusiast's efficiency and success in the field.

Furthermore, the continuous evolution of metal detector technology is a testament to the enduring human fascination with discovery and exploration. As new advancements are made, enthusiasts will find it increasingly easier and more rewarding to uncover hidden treasures that tell the rich stories of our past. Thus, the history of metal detectors is not just a tale of technological advancement but also a key to unlocking the countless mysteries buried beneath our feet.

Types of Metal Detectors

When exploring the realm of metal detecting, the type of metal detector you choose can dramatically influence your success and enjoyment of the hobby. Each type of detector has its own specific set of technologies and features, designed to cater to different conditions and targets.

Very Low Frequency (VLF) detectors are among the most common and versatile. They operate using two coils, one to transmit and one to receive, which work together to detect metallic objects and differentiate between various types of metals. This differentiation is crucial for those who wish to filter out trash metal and focus on more valuable finds like coins, jewelry, and relics. VLF detectors are particularly effective in areas with minimal ground mineralization and are favored for their ability to provide a balance between depth and sensitivity.

Pulse Induction (PI) detectors, on the other hand, use a single coil as both transmitter and receiver or have two or three coils working together. They send powerful, brief magnetic pulses into the ground, which are ideal for environments highly mineralized such as saltwater beaches or gold-rich terrains where traditional VLF detectors might struggle. PI detectors excel in ignoring ground minerals and can find deeper buried objects, but they generally lack the ability to discriminate between different types of metals as effectively as VLF detectors.

Beat Frequency Oscillation (BFO) detectors are the simplest and least expensive type. They utilize two coils, each with its own oscillator that creates a frequency. The frequencies from the two oscillators interfere with each other, and the detector uses this interference to identify metal. While BFO detectors are less sophisticated and offer more limited depth and sensitivity than VLF or PI models, they provide an excellent entry point for beginners due to their ease of use and affordability.

Choosing the right type of metal detector involves considering the specific environments where you plan to search and what you are most interested in finding. While VLF detectors offer great all-around utility for a variety of conditions, PI models are better suited to challenging environments and BFO detectors can offer a simple, cost-effective start to your metal detecting journey. Understanding these types helps users not only in selecting the right equipment but also in mastering its use for the best possible outcomes. Each technology has its own learning curve, and knowing how to adjust the settings to suit specific hunting conditions can be the key to success. This knowledge ensures that the time spent in fields, beaches, or old homesteads is both productive and enjoyable.

Chapter: 1 Understanding Metal Detector Technology

Basic Operating Principles

Understanding the basic operating principles of metal detectors is essential for anyone interested in using this technology effectively. At its core, a metal detector is a tool that uses electromagnetic fields to locate metal objects buried underground or hidden in other objects.

The primary mechanism of a metal detector involves generating an electromagnetic field from its search coil into the surrounding environment. When this field comes into contact with a metallic object, the metal's conductive properties disturb the field, and the disturbance is detected by the receiver coil in the detector.

This disturbance alters the original electromagnetic field's frequency or phase. Modern metal detectors are designed to interpret these changes to determine not only the presence of metal but also provide clues about the metal's identity and depth. This process involves a complex interplay of physics and electronics that can be broken down into several key steps:

1. Transmission of Electromagnetic Field: When the metal detector is activated, the search coil transmits a steady electromagnetic field into the ground. This field is generated by currents flowing through the coil, creating a magnetic field that extends outward from the coil into the surrounding area.

2. Interaction with Metal Objects: As the electromagnetic field permeates the ground, any metal objects in the vicinity become energized due to their conductive properties. This interaction induces eddy currents in the metal object, which in turn generate their own magnetic fields. It's these fields that the detector senses.

3. Reception of Signal: The secondary magnetic field created by the eddy currents in the metal object interacts with the receiver coil in the detector. This interaction alters the phase and amplitude of the voltage in the coil. The metal detector's electronics are tuned to detect these subtle changes and analyze them based on the coil's original signal.

4. Signal Processing: The signal received by the detector is processed to filter out noise and improve the clarity of the detection signal. This processing helps differentiate between different types of metals and can often give an indication of the depth at which the object lies. Advanced metal detectors use sophisticated digital processing algorithms to enhance detection and discrimination capabilities.

5. Readout: Once the signal has been processed, the metal detector provides feedback to the user, usually via audio tones, visual indicators, or a combination of both. Different metals will produce distinct responses, and experienced users can often identify the type of metal detected based on the signal characteristics.

By mastering these basic operating principles, users can optimize their search strategies, adjust their equipment settings appropriately for different environments, and significantly increase their chances of finding valuable metal objects. Whether searching for relics, jewelry, coins, or other metals, understanding how a metal detector works is the first step towards successful metal detecting. This knowledge not only enhances the effectiveness of the search but also adds to the enjoyment and satisfaction of uncovering hidden treasures.

Types of Detection Technology

Metal detectors come equipped with various technologies, each designed to enhance the detection capabilities and efficiency depending on the environment and type of metal being sought. These technologies primarily involve differences in how detectors transmit and receive electromagnetic fields to identify metal objects. Understanding these technologies is crucial for users to choose the right equipment and optimize their metal detecting strategies.

The most common technology used in metal detectors is Very Low Frequency (VLF) technology. This system employs two distinct coils: a transmitter coil and a receiver coil. The transmitter coil generates a magnetic field that, when disrupted by a metallic object, induces a signal in the receiver coil. This change is detected and analyzed by the detector's control box. VLF detectors are particularly adept at distinguishing between different types of metals, which allows users to set their devices to ignore unwanted materials, such as iron. This capability makes VLF technology highly suitable for coin shooting, relic hunting, and general treasure hunting in relatively mineral-free soil.

Another sophisticated technology is Pulse Induction (PI). Unlike VLF, which uses continuous electromagnetic transmission, PI technology sends powerful, brief bursts of current through its coil or coils. These pulses generate a short magnetic field that, when it collapses suddenly, sends back an echo detected by the sensor circuitry in the detector. PI systems can penetrate highly mineralized soil, saltwater, and other challenging environments where VLF detectors might struggle. They are less sensitive to ground mineralization and can detect objects buried at greater depths. However, their major drawback is the lack of discrimination capabilities compared to VLF, as they pick up all metal types without differentiation.

Beat Frequency Oscillation (BFO) technology is the simplest form of metal detection technology and tends to be found in less expensive detectors. BFO detectors use two coils, each with an oscillator that creates a frequency. These frequencies interact when a metal object is present, altering the frequency at the detector's audio output. While BFO detectors are less accurate and have poorer depth and sensitivity than VLF or PI detectors, they are an excellent option for beginners due to their simplicity and cost-effectiveness.

For users, choosing the right technology depends on several factors, including the typical metal types they aim to find, the environments in which they will be searching, and how much they are willing to invest. VLF is suitable for beginner to advanced users who need discrimination features and are primarily searching lands with low to moderate soil mineralization. PI technology suits those working in highly mineralized environments, such as beaches or gold-rich areas, who need powerful detection capabilities and are less concerned with sorting trash from treasure. BFO is best for casual hobbyists or those just beginning to explore metal detecting.

In practice, knowing how to use these technologies involves adjusting settings like sensitivity, discrimination, and ground balance, which can vary significantly across different models and brands. Each type of technology requires a different approach to handling background mineralization and junk metals, and mastering these settings can greatly enhance the effectiveness and satisfaction of metal detecting adventures.

Key Components of a Metal Detector

Understanding the key components of a metal detector is crucial for anyone looking to get the most out of this fascinating technology. Each part of the detector plays a specific role in the process of locating and identifying buried metals, making it essential to grasp how these components work together to achieve successful results.

At the heart of every metal detector is the control box, which houses the circuitry, controls, speaker, batteries, and microprocessor. This component acts as the brain of the detector, processing the signals that are picked up by the search coil, distinguishing between different types of metals, and providing feedback to the user. The control box allows for adjustments to sensitivity, discrimination, and other operational modes, making it possible to tailor the detector's performance to specific conditions and targets.

The search coil, also known as the detector loop or search head, is arguably the most critical part of the metal detector. Available in various sizes and shapes, coils send out an electromagnetic field into the ground and detect the return signal from metallic objects. The size and shape of the coil affect the depth and scope of the search; larger coils cover more ground and detect deeper, whereas smaller coils provide greater sensitivity and precision for smaller targets. Coils are specifically designed to be interchangeable to suit different searching environments.

Attached to the control box and the coil is the shaft, which connects all the components together and allows for adjustments in length to suit the height and comfort of the user. This adjustability is important as it ensures that the detector can be used comfortably for extended periods, reducing fatigue and increasing efficiency.

The stabilizer, often overlooked, is another vital component. It is usually a padded arm brace that allows the detector to rest against the user's arm, providing stability during the sweeping motion. This part helps reduce strain on the arm and improves control over the sweeping motion of the coil, leading to more accurate detecting.

Understanding these components and their functions not only aids in selecting the right metal detector but also enhances the user's ability to effectively use the device. Knowing how to adjust the control settings, choose the appropriate coil for the terrain, and maintain proper posture with the shaft and stabilizer can significantly improve the detection process and increase the likelihood of finding valuable metals hidden beneath the surface. Each part of the metal detector is essential in turning an ordinary outdoor walk into a treasure hunt.

Chapter: 2 Setting Up Your Metal Detector

Assembly Instructions

Setting up your metal detector correctly is essential for ensuring its optimal performance and enhancing your metal detecting experience. The process of assembling a metal detector might vary slightly depending on the brand and model, but most share common steps that are easy to follow even for beginners.

Begin by unpacking all the components from the box. You should have a control box, which is the brain of the detector housing the electronics and user interface; a search coil, which actually detects the metal; a shaft, which connects the coil to the control box and is usually adjustable to suit your height; and a stabilizer (sometimes called an armrest) to keep the detector steady as you sweep.

First, attach the search coil to the lower end of the shaft. This is typically done by sliding the coil onto the shaft and securing it with a set of included washers and a knob or screw. Ensure this connection is tight to prevent wobbling, but be careful not to overtighten as this can strip the threads or damage the coil.

Next, connect the shaft to the control box. This step usually involves sliding the upper end of the shaft into a slot or bracket on the control box and securing it with a locking mechanism. In some models, you might find that the control box first needs to be attached to the stabilizer or armrest, which then connects to the shaft. Adjust the length of the shaft so that when you stand upright, the coil rests flat on the ground while your arm rests comfortably on the stabilizer. This will help you avoid fatigue during long sessions of metal detecting.

After the physical assembly, you will need to connect the search coil's cable to the control box. This cable transmits the signals from the coil to the brain of the detector for analysis. Run the cable along the shaft and secure it at several points using the clips or Velcro straps provided. This prevents the cable from snagging or getting tangled while in use. Connect the cable to the designated port on the control box, making sure the connection is secure but not forced.

Once the physical setup is complete, it's time to power on your device. Insert the batteries in the battery compartment, which is usually on the control box. Some detectors also have a battery pack that is worn around the waist or slung over the shoulder, which connects to the control box via a cable.

With the metal detector assembled and powered, turn it on and perform an initial test to ensure everything is working as expected.

This is a good time to familiarize yourself with the controls and default settings. Adjust the settings according to the conditions you expect to encounter, such as ground balance, sensitivity, and discrimination, based on your location and target metals.

Now that your metal detector is fully assembled and tested, you're ready to start hunting for treasures. Remember, the initial setup is crucial for a smooth and efficient detecting experience. Proper assembly not only helps in maintaining the longevity of the device but also enhances its functionality, making your adventures in metal detecting both productive and enjoyable.

Initial Calibration and Settings

Setting up a metal detector correctly through initial calibration and settings is crucial to ensuring optimal performance during your treasure hunting endeavors. The process involves a series of adjustments and tests that tune the detector to the specific conditions of the search area, as well as to the user's preferences for what types of targets they wish to find.

The first step in the initial setup is the assembly of the metal detector, which usually includes attaching the search coil to the lower shaft, connecting the shaft to the control box, and ensuring all cables are securely fastened to prevent false signals. Once assembled, the real task of calibration begins.

Powering up the metal detector is typically followed by a basic calibration known as the 'ground balance' adjustment. Ground balancing is vital because it allows the detector to compensate for minerals in the soil, reducing the likelihood of false positives created by naturally occurring metal components in the ground. Manual ground balance involves holding the coil just above the ground, pressing a button to initiate the balance, and then lifting and lowering the coil several times until the detector stops giving feedback from the ground. Some advanced detectors offer automatic or pre-set ground balance settings that adjust based on the detected ground conditions, making it easier and quicker to start searching.

Sensitivity settings are another critical adjustment. Sensitivity determines how deep the detector scans and how small the objects are that it can detect. Setting the sensitivity too high might lead to constant false signals in highly mineralized soils, while setting it too low might cause you to miss smaller or deeper objects. It's often recommended to start with a medium sensitivity setting and adjust according to the frequency of signals and the types of ground being searched. More sophisticated metal detectors will have presets and automatic adjustments to help find the right balance.

The discrimination setting is equally important, as it helps the detector ignore unwanted objects like pull-tabs, nails, and other metallic trash. By setting the discrimination level, the detector can ignore specific metal types or signal ranges. This setting is crucial in areas where there is likely to be a lot of garbage. However, setting discrimination too high can also cause you to miss valuable items that fall within the blocked signal range, so careful adjustment is necessary.

For metal detectors with multiple operation modes, selecting the correct mode for the intended type of hunting is part of the initial setup. Modes such as 'coins', 'jewelry', 'beach', and 'relic' can preconfigure balance, sensitivity, and discrimination settings to optimize the detector for those specific conditions and targets.

Testing the setup in a controlled environment, like a garden or a test bed with different metal objects buried at known depths, can help fine-tune these settings. Observing how the detector reacts to different metals and objects can provide valuable insights and allow further tweaking to improve accuracy and efficiency.

In summary, the initial calibration and settings of a metal detector involve setting up the ground balance to negate responses from mineralized soil, adjusting sensitivity to detect desired object sizes at appropriate depths, setting discrimination to ignore unwanted metals, and choosing the right mode for the task at hand. These steps are foundational for anyone looking to get the most out of their metal detecting experience, making them crucial elements in the preparation phase of any metal detecting activity.

Understanding and Adjusting Sensitivity, Discrimination, and Ground Balance

Understanding and adjusting the sensitivity, discrimination, and ground balance of a metal detector are essential skills for any metal detecting enthusiast. These settings play a pivotal role in optimizing the detector's performance and enhancing the user's ability to find valuable items while minimizing junk finds.

Sensitivity in a metal detector determines how much signal is picked up from metal objects in the ground. Higher sensitivity settings increase the detector's ability to detect smaller or deeper objects, which can be crucial when searching in areas where valuable items are buried deep. However, setting the sensitivity too high can lead to false signals and chatter from mineralized soils, wet beach sand, or nearby metal trash. The key is to find a balance where the sensitivity is high enough to detect the objects you are interested in, but not so high that you are overwhelmed with false signals. Beginners may want to start with a lower sensitivity to better learn their machine's feedback and gradually increase as they become more comfortable with the sounds and signals.

Discrimination is a function that allows the detector to ignore certain types of metal based on their conductivity. By setting the discrimination level, users can prevent the detector from signaling every metal object, such as iron or foil, thus focusing on more desirable targets like coins or jewelry. The discrimination setting is particularly useful in areas with a lot of trash, as it helps save time and effort digging up undesirable items. However, overly high discrimination can also lead to skipping over valuable but low-conductive items such as gold, so careful adjustment is crucial. It's important to experiment with different discrimination settings in various locations to understand how it affects the detection of different metals.

Ground Balance is a feature that allows the metal detector to ignore the effects of minerals in the soil. In highly mineralized soils, failing to properly ground balance can result in poor depth and false signals. Manual ground balancing allows the user to adjust the detector so that it does not react to the mineral content in the ground. Some advanced detectors offer automatic ground balancing, which adjusts to the soil conditions dynamically as you detect, which can be a great feature for beginners or when working in varied soil conditions. Proper ground balancing improves the depth and sensitivity of the detector, providing a more accurate identification of buried objects.

Mastering these settings requires a combination of reading the metal detector's manual, understanding the theory behind metal detection, and a lot of practice in the field. Experimenting with different settings in a controlled environment, such as a test garden where various types of metals are buried at known depths, can be extremely helpful. It allows the user to see how changes in settings affect the types of metal detected and the depth at which they are detected. This practice builds a practical understanding of how to set up a metal detector optimally for any given situation, improving both the efficiency and enjoyment of the detecting experience. As these skills develop, the detector becomes not just a tool, but an extension of the user, capable of unlocking secrets hidden beneath the ground.

Chapter: 3 Techniques for Successful Metal Detecting

Searching Techniques

Mastering effective searching techniques is essential for successful metal detecting, enhancing both the quantity and quality of finds. Understanding the ground, handling the equipment correctly, and employing systematic search patterns are key components that can make a significant difference in outcomes.

Starting with the ground sweep, the technique with which a detectorist handles the coil is foundational. The coil should be kept level and close to the ground at all times to maximize depth and sensitivity. Sweeping the coil in a slow, controlled manner from side to side, slightly overlapping each sweep, ensures comprehensive coverage of the area. This technique prevents missed spots and increases the likelihood of detecting smaller or deeper objects that might otherwise be overlooked.

Ground coverage is further enhanced by adopting a methodical search pattern. The most common pattern is a grid approach, where the area is divided into manageable sections. Detectorists often start at one corner and make parallel passes back and forth across the section, then move to the next section and repeat the process. This organized method ensures that every inch of the search area is thoroughly scanned, and it helps in keeping track of which areas have been covered.

Pinpointing the exact location of a detected object is another critical technique. Most modern metal detectors are equipped with a pinpoint function that narrows the search area when a potential target is detected. By using this feature, detectorists can minimize the size of the hole they need to dig, thereby reducing the impact on the environment and saving time. The pinpointing process usually involves moving the coil in small circles or side-to-side movements over the target area until the strongest signal is found, indicating the exact location of the object.

The depth of a target can often be gauged by the strength of the signal and the settings of the metal detector. Learning to interpret these signals is a skill developed over time. Some detectors provide an estimate of depth on the display, which can guide the user on how deep to dig.

Another advanced technique involves understanding and adapting to different soil types and environmental conditions. For example, wet sand conducts electricity better than dry sand, enhancing the detector's ability to find metal objects. Similarly, colder temperatures can sometimes reduce battery performance, affecting the detector's sensitivity. Adapting the detector's settings to match these conditions, such as adjusting the ground balance to cope with highly mineralized soils, can significantly improve the success rate.

Finally, detectorists should remain aware of the legal and ethical considerations involved in metal detecting. Always obtaining permission to search on private property, adhering to local laws regarding historical sites, and practicing "Leave No Trace" principles by properly filling any holes dug and disposing of any trash encountered are practices that preserve the hobby's integrity and sustainability.

By employing these techniques, those who use metal detectors can significantly increase their effectiveness, resulting in more rewarding and responsible treasure hunting experiences. Each session offers a learning opportunity, and over time, these techniques become second nature, enhancing the enjoyment and success of metal detecting adventures.

Target Identification

In the context of using a metal detector, mastering target identification is crucial for distinguishing valuable finds from common metal trash. This skill hinges on interpreting the feedback from your metal detector effectively and knowing what to do with the information it provides.

When a metal detector locates a potential target, it typically signals the find with an audio tone or a visual cue on its display, if equipped. Understanding these signals is the first step in target identification. Higher-end detectors often include target identification (ID) features that can indicate probable metal types based on the conductivity of the detected object. These IDs are usually displayed as numerical values or specific icons that represent different metals such as iron, silver, gold, or nickel.

To become proficient in target identification, a user must spend time learning the nuances of their detector's feedback. This involves not only recognizing the different sounds or visual signals for various metals but also understanding the subtleties in the tone or pitch that might distinguish a valuable coin from a piece of foil or a nail. For instance, a consistent, high-pitched tone or a specific number range on the detector's digital readout can suggest a non-ferrous metal like copper or silver, which is often more desirable.

Ground balancing is another technique that enhances target identification. This feature allows the detector to ignore signals from mineralized soils that can mask or mimic metal targets. By adjusting the ground balance, the detector becomes more sensitive to the metals buried in the ground and less likely to provide false positives.

Additionally, pinpointing is a technique used to locate the exact position of the detected item more accurately. Most detectors have a pinpoint mode, which helps refine the search area once a potential target is indicated. Using this mode reduces the size of the area to be dug, minimizes the disturbance to the surrounding ground, and increases the likelihood of retrieving the target efficiently.

Experience also plays a significant role in target identification. Over time, detectorists learn to interpret faint signals and differentiate between overlapping responses, which is common in areas where multiple items lie close together. Experienced users can often tell just by the quality of the audio tones whether a target is worth digging.

Practicing in different environments also aids in honing identification skills. For instance, beaches, parks, and old homesites each present unique challenges and opportunities for learning the characteristics of various metals under different conditions. Furthermore, engaging with the metal detecting community, whether through online forums, clubs, or group hunts, can provide insights and tips from more experienced detectorists, enhancing one's own ability to identify targets accurately.

Ultimately, effective target identification reduces time and effort spent on undesirable finds and increases the overall success and satisfaction of the treasure hunting experience. Each outing with a metal detector can refine a user's skills, gradually building their proficiency in navigating through the myriad of signals to uncover the treasures hidden beneath the surface.

Digging and Retrieving Items

Digging and retrieving items is a crucial aspect of metal detecting that requires careful technique to preserve both the found item and the surrounding environment. Successful retrieval begins with identifying a potential target through your detector's signals. Once a target is pinpointed, the process of excavation should be carried out with precision and respect for the location.

The first step in digging for an item is to use a handheld pinpointing metal detector if available. This device helps refine the target area so that digging is more accurate, reducing the amount of ground disturbed. Once the exact location is identified, the type of tool used for digging becomes important. For softer ground, a standard garden trowel may suffice, but for harder, more compact soils, a specialized digging tool known as a digging knife or a lesche digger is more appropriate. These tools are designed to cut into the ground with minimal disruption.

The technique for cutting into the soil is also crucial. One popular method is to cut a 'plug'—a U-shaped flap of turf or soil. This involves inserting the digger around the target and lifting out the plug, which can then be replaced without leaving permanent damage to the area. It's important to ensure that the grass or turf on top of the plug remains intact and undisturbed as much as possible. For environments where grass or delicate ecosystems are present, the plug technique is preferred to maintain the natural aesthetics and health of the area.

Once the plug is removed, you can use either your hands or a smaller tool to retrieve the item from the loosened soil within the hole. Care should be taken to disturb as little of the surrounding earth as possible. This not only makes it easier to close the excavation neatly but also minimizes the impact on the environment.

After retrieving the item, inspecting it, and cleaning off any loose soil, it's important to properly fill in the hole. Replace the plug or soil exactly how it was extracted, pressing it down firmly to ensure it settles back into place seamlessly. If in a highly visible or trafficked area, additional care should be taken to leave no trace of digging.

Beyond the immediate technique, responsible metal detecting includes carrying away any trash that might be unearthed during the search. This practice helps clean up the environment and promotes a positive image of metal detecting to the public and landowners.

Lastly, after retrieving and inspecting the find, careful cleaning and preservation are crucial, especially if the item is old or potentially valuable. Different materials require different cleaning techniques; for instance, old coins might need gentle brushing or professional cleaning to avoid damage, whereas modern items made from sturdier materials might simply need a wash under warm water.

By mastering these digging and retrieving techniques, metal detector enthusiasts not only increase their chances of a successful find but also ensure that their hobby can be enjoyed sustainably and respectfully for years to come. This thoughtful approach helps in preserving the historical and environmental integrity of the locations they explore.

Metal Detecting Etiquette

Metal detecting is not just a solitary pursuit of hidden treasures; it's also a hobby that requires adherents to follow a set of unspoken rules or etiquette to ensure that the activity remains enjoyable and sustainable for everyone involved. Good metal detecting etiquette involves respecting both the environment and other people, whether they are fellow hobbyists, property owners, or passersby.

First and foremost, always obtain permission before detecting on private land. This is crucial, as trespassing can not only lead to legal troubles but can also tarnish the reputation of the metal detecting community. For public lands, it's important to familiarize yourself with local laws and regulations. Some areas might be off-limits for metal detecting, while others may require a permit. By respecting these rules, detectorists help preserve historical and ecological sites and ensure that these areas remain accessible for future detecting.

When you do start detecting, it is vital to minimize your impact on the land. This means practicing careful recovery methods. When digging for objects, use appropriate tools such as a narrow digging tool or a probe. Make a neat plug (a small, U-shaped flap of sod) that can be easily replaced without leaving noticeable signs of disturbance. After recovering an item, replace the plug and any dislodged dirt or sand, and press the earth firmly back into place. The goal is to leave the site looking as untouched as possible.

Another aspect of metal detecting etiquette is to be considerate of others around you. If you're detecting in a public space like a beach or park, be mindful of your surroundings. Keep the noise of your detector headphones to yourself, maintain a respectful distance from sunbathers and picnickers, and avoid blocking pathways or recreational areas. The presence of a metal detector should not be a nuisance to others.

Responsible disposal of findings is another key element of good etiquette. Always take away any trash that you dig up, such as aluminum foil, nails, and can tabs. Not only does this clean up the environment, making it safer and more pleasant for others, but it also enhances the reputation of metal detecting as a beneficial activity.

Engaging with the community is also part of metal detecting etiquette. Sharing findings and tips with other detectorists can help foster a sense of camaraderie and support among enthusiasts. Moreover, participating in forums, clubs, or online communities can provide valuable insights into best practices and new technologies, as well as updates on legislation affecting metal detecting.

Lastly, always respect historical artifacts. If you find something that seems to be of historical significance, it's important to report it to local authorities or historical societies. In many places, removing historical artifacts without permission is illegal, and doing so can result in serious penalties. Respecting these items ensures they can be studied and preserved for public benefit and historical research.

By adhering to these guidelines, metal detectorists can ensure that their hobby not only brings them personal satisfaction but also contributes positively to the community and the environment. Good etiquette is about fostering a sustainable practice that respects history, nature, and fellow enthusiasts, ensuring that metal detecting can be enjoyed by generations to come.

Chapter: 4 Choosing the Right Equipment

Selecting the Best Metal Detector for Your Needs

Selecting the best metal detector for your needs is a critical decision that can significantly affect your success and enjoyment in metal detecting. This process involves understanding the different types of detectors available, assessing your main hunting environments, and considering your budget and personal interests.

First, consider the primary location where you will be using your metal detector. Different environments such as beaches, parks, old battlefields, or near riverbanks can drastically influence the type of detector you should choose. For example, if you're planning to search primarily on beaches, look for a metal detector specifically designed for saltwater. These models often feature pulse induction technology or are very low frequency (VLF) detectors with adjustable ground balance settings that help manage the highly mineralized ground conditions of wet sand.

Second, think about what you are most interested in finding. Metal detectors come with features that can be tuned to optimize the search for specific types of metal like gold, silver, coins, or relics. Detectors with a higher frequency are generally better at finding small pieces of gold, which is particularly useful if you're prospecting for gold flakes or nuggets. If your focus is on coins, jewelry, or relics, a detector with good discrimination features and the ability to filter out junk metal would be ideal.

Third, evaluate the detector's features in terms of your experience level. For beginners, it's advisable to start with a simpler, user-friendly model that does not require complex adjustments. Features such as preset ground balance, straightforward discrimination controls, and an intuitive user interface will make the initial learning process less daunting. More experienced users, however, might prefer detectors that offer more advanced settings like manual ground balance, adjustable thresholds, and enhanced sensitivity settings, which allow for more precise tuning in varied conditions.

Fourth, consider the ergonomics and durability of the metal detector. Since metal detecting often involves several hours of physical activity, the weight and design of the detector should be comfortable for extended use. Look for a lightweight model with an adjustable handle and armrest to reduce strain on your arm and back. Durability is also crucial, especially if you will be detecting in rough terrain or extreme weather conditions; thus, a weather-resistant and rugged design can be a wise investment.

Finally, your budget will play a significant role in your decision. Metal detectors can range from under a hundred dollars to several thousand, based on their capabilities and features. It's important to balance cost with the specific features you need—investing in a more expensive detector makes sense if you are committed to the hobby and require more advanced features. However, for occasional use or beginning in the hobby, a mid-range detector might be the most cost-effective and practical choice.

In conclusion, selecting the right metal detector is about matching the machine's capabilities and features with your specific needs and goals. By carefully considering the environment, target types, your experience level, the detector's ergonomics and durability, and your budget, you can choose a metal detector that will enhance your searching experience, leading to more successful and enjoyable hunts.

Essential Accessories for Metal Detecting

For anyone delving into the world of metal detecting, understanding the significance of essential accessories is key to enhancing the overall experience and efficiency of your hunts. These accessories not only improve the functionality of your metal detector but also contribute to the care and longevity of your equipment and finds.

Headphones are one of the most important accessories for metal detecting. They allow you to hear subtle changes in tone that are crucial for identifying what lies beneath the surface, especially in noisy environments. Quality headphones can block external noise and help you focus on detecting, thereby increasing your chances of finding smaller or deeper objects. They can also extend the battery life of your metal detector by reducing the need for a loudspeaker. For optimal experience, look for headphones that are comfortable, provide clear sound, and are compatible with your metal detector's audio output.

Digging tools are essential for retrieving your finds without damaging them or the surrounding area. The type of tool you choose often depends on the terrain you are searching. For soft soil, a hand trowel or digging knife might suffice, while a heavier duty shovel may be necessary for tougher grounds like clay or rocky soil. Many experienced detectorists carry a variety of tools, including a sand scoop for beach hunting. It's important to select tools made from durable materials that resist rust and wear. Also, consider tools with a serrated edge for easier soil penetration and features like a ruler to measure the depth of holes.

Carrying cases are crucial for protecting your metal detector and accessories during transport and storage. These cases can prevent your equipment from being jostled or damaged and help organize your gear, which is particularly handy when traveling to remote locations. A good carrying case will be sturdy, lightweight, and designed to fit your specific model of metal detector along with compartments for additional accessories like coils, batteries, and headphones.

Cleaning kits are indispensable for maintaining both your metal detector and the items you find. Regular cleaning of your metal detector is important to prevent dirt, sand, and water from affecting its performance. The kit should include brushes, cloths, and cleaners that are safe to use on electronic components. For the items you unearth, especially historical artifacts or coins, having gentle cleaning tools can help preserve their condition and value. This often includes soft brushes, wooden picks, and mild cleaning solutions.

Incorporating these essential accessories into your metal detecting practice not only enhances your ability to find treasures but also helps in the maintenance and longevity of your equipment. Each accessory plays a pivotal role in the success of your treasure hunting endeavors, from the moment of discovery to the care and handling of your finds. Choosing the right equipment is about balancing functionality with comfort and preservation, ensuring every metal detecting session is productive and enjoyable.

Chapter: 5 Maintenance and Care

Routine Maintenance Tips

Routine maintenance of a metal detector is essential for ensuring its longevity and reliability. Proper care not only enhances the detector's functionality but also prevents malfunctions that could impede a fruitful treasure hunting experience. Here are some comprehensive tips on maintaining your metal detector to keep it in optimal condition.

First and foremost, always clean your metal detector after each use. Dirt, mud, sand, and salt can accumulate on the coil and in the crevices of the device, which may lead to corrosion or damage if left unchecked. Use a soft brush to remove debris from the coil and a damp cloth to wipe down the control box and shaft. Avoid using harsh chemicals or abrasives, as these can damage the detector's finish and potentially interfere with its electronic components.

It's also important to prevent water damage, especially if your detector is not fully waterproof. While many modern detectors have waterproof coils, their control boxes may not withstand exposure to moisture. If you are detecting in damp environments or near water, ensure that any non-waterproof components are adequately protected with a cover or case. After detecting in humid or wet conditions, open up any compartments (like the battery compartment) to air them out and dry any condensation or moisture that might have accumulated.

Another key aspect of maintenance is regularly checking and tightening the screws and fittings on your metal detector. The constant motion of swinging the detector back and forth can loosen fittings over time, which could lead to parts wobbling or becoming detached. Tighten these components gently, being careful not to strip the screws, which can happen if too much force is applied.

Battery care is crucial, as improper battery maintenance is a common cause of detector failure. Always remove batteries before storing your detector, particularly if you do not plan to use it for an extended period. This prevents the batteries from corroding inside the device, which can cause irreversible damage. Additionally, use only the recommended types of batteries, as some detectors require specific voltages or types of batteries to operate efficiently.

Software updates are often overlooked but are vital for digital detectors that rely on firmware. Manufacturers frequently release updates that enhance features, improve functionality, and fix bugs. Keeping your detector's software up to date ensures that you are utilizing all its capabilities and that it operates smoothly.

Finally, consider storing your metal detector in a cool, dry place away from direct sunlight and extreme temperatures, which can degrade its electronic components over time. Use a padded bag or a hard case to protect it from physical damage when not in use. Proper storage not only prevents wear and tear but also keeps the detector safe from accidental drops or impacts that could misalign its components or damage its internal electronics.

By following these routine maintenance tips, you can extend the life of your metal detector and ensure it remains a reliable tool for your treasure-hunting adventures. Regular care not only preserves the physical and operational integrity of your detector but also enhances your overall experience by minimizing the likelihood of malfunction during use.

Troubleshooting Common Problems

In the world of metal detecting, maintaining and troubleshooting your equipment is crucial for ensuring a successful and enjoyable treasure hunting experience. Regular maintenance and understanding how to troubleshoot common problems can extend the life of your metal detector and enhance its performance.

One common issue that many detectorists encounter is inconsistent signals or false alarms. This can often be attributed to low battery power. It's essential to start each hunting session with fully charged or new batteries to ensure optimal performance. Also, false signals can occur when the sensitivity setting is too high for the conditions. Adjusting the sensitivity lower until the detector stabilizes can often resolve this issue.

Another frequent problem is the detector not turning on. This might seem basic, but it's always worth checking to make sure that the batteries are correctly installed and that the battery terminals are clean. Corrosion on the terminals can be cleaned with a bit of vinegar or lemon juice to ensure a good connection. If the detector still fails to turn on after these checks, it could indicate a deeper electrical issue, such as faulty wiring or a broken power switch, which might require professional repair.

Ground balancing issues are also common, particularly for those detecting in highly mineralized soils. If a detector is not properly ground balanced, it may give erratic readings or miss targets altogether. Many modern detectors have automatic ground balancing, but manual balancing can often yield better results. Reviewing the manufacturer's instructions to correctly perform a ground balance adjustment according to your specific model is essential.

Cable and connector issues can lead to unstable operation and erratic signals. Regularly inspect the coil cable for any signs of wear or damage. Ensure that the cable is securely connected and that there is no undue tension pulling on the connections. Loose or damaged cables may need to be replaced to restore proper functionality.

Coil problems are another area of concern. Damage to the coil can occur through regular use, particularly if you frequently detect in rugged terrains. Inspect the coil for cracks or splits, which can compromise its effectiveness. Also, ensure that the coil is securely attached; a loose coil can lead to poor performance. Sometimes, problems perceived as being internal to the detector can be resolved simply by replacing or securely attaching the coil.

Interference from external sources can also impact the performance of your metal detector. Power lines, other metal detectors, and even cell phones can cause erratic signals. If you suspect interference, try adjusting the frequency on your detector if it has that capability or move to a different location to see if the problem persists.

Finally, regular cleaning and storage are vital. After each use, clean the detector, especially the coil, with a soft cloth to remove dirt and debris. Never use harsh chemicals or abrasive cleaners. Store the detector in a dry, cool place out of direct sunlight. Avoid storing the detector with the batteries inside, as they can leak and cause damage over time.

By becoming familiar with these common issues and learning how to quickly address them, you'll keep your metal detecting hunts productive and your equipment in top condition. Keeping the manual handy for troubleshooting specific to your model can also save time and frustration. This proactive approach to maintenance and care not only prolongs the life of your metal detector but also enhances your overall experience with the hobby.

Long-term Storage Solutions

Proper long-term storage of a metal detector is crucial to maintaining its functionality and extending its lifespan. As these devices consist of sensitive electronic components and metallic parts, they require careful handling and specific conditions to prevent damage while not in use.

Firstly, cleaning the metal detector before storage is essential. Remove any dirt, sand, or mud from the coil and shaft, using a soft brush or cloth to avoid scratching any surfaces. Ensure the control box, which houses the electronics, is wiped down gently. If the model is not fully waterproof, extra care should be taken to avoid moisture ingress in this area.

Dismantling the metal detector can further protect it during storage. Most detectors come with a breakdown feature, allowing the shaft to be separated into smaller parts and the coil to be detached. Breaking down the unit minimizes the required storage space and reduces the risk of accidental bending or pressure on any parts.

For storage locations, choose a cool, dry place away from direct sunlight, extreme temperatures, or high humidity, which can degrade electronic components and cause metal corrosion. Basements and garages are common storage areas but can be prone to dampness and temperature fluctuations, so ensure the environment is controlled or that the detector is adequately shielded in such conditions.

Battery care is another crucial aspect. Always remove batteries from the detector to prevent leaks which can corrode and permanently damage the electronic circuitry. Store the batteries separately in a cool, dry place and check them periodically for signs of leakage or corrosion if they are reusable.

Additionally, it's advisable to invest in a protective case or cover for the metal detector. This cover can prevent dust buildup and shield the detector from environmental factors like humidity and temperature swings. If a professional case isn't available, wrapping the detector in a soft, breathable material can suffice.

Lastly, for those who store their detectors for very long periods, periodic checks every few months are recommended. This involves assembling the detector, turning it on to check its functionality, and possibly sweeping it over a metal object to ensure it still detects correctly. This practice not only ensures the detector remains operational but also helps identify potential issues that could worsen over time.

By adhering to these guidelines, enthusiasts can ensure their metal detectors are ready and fully functional for the next treasure hunting adventure, even after long periods of storage. This attention to proper care and maintenance will prolong the effective life of the detector, ensuring many successful searches in the future.

Chapter: 6 Legal and Ethical Considerations

Understanding Local Laws and Regulations

Understanding local laws and regulations is a critical component of responsible metal detecting. Metal detector enthusiasts must navigate a variety of legal landscapes that can vary significantly from one location to another. It is essential to respect these laws to avoid legal penalties and to maintain good relationships within the community.

In the United States, metal detecting is subject to federal, state, and local regulations that can affect where and what you can search for. On federal lands, for example, metal detecting is often strictly prohibited or heavily regulated. National parks, for instance, generally do not allow metal detecting as they are protected areas where the disturbance of the soil is not permitted. This protection extends to historical artifacts, which are preserved for their cultural and historical significance.

State laws can also vary. Some states allow metal detecting in state parks with specific stipulations, such as requiring a permit or restricting detecting to certain areas. It's common for these regulations to include a stipulation that any significant finds must be reported to park authorities. This is particularly true for items that could be considered historical artifacts. Some state beaches, however, may be more lenient, allowing detecting without specific permits but still enforcing rules against digging in certain areas such as dunes, which are often protected for environmental reasons.

Local ordinances must also be considered. Many cities and towns have their own rules regarding metal detecting, which might restrict the use of metal detectors in public parks or municipal properties without explicit permission. In some places, metal detecting might be allowed year-round, while in others it might be restricted during busy seasons or in designated areas only.

Ethically, it is important to always seek permission where necessary. If you're interested in detecting on private property, obtaining explicit permission from the property owner is a must. This not only avoids legal complications but also helps in building trust and respect within the community.

Additionally, metal detecting is not just about following legal guidelines but also about adhering to a code of ethics. This includes restoring the ground to its original condition after digging, properly disposing of any trash uncovered, and being considerate of others enjoying the outdoor spaces.

Responsible metal detecting means staying informed about the laws and regulations in each new area you explore. Many metal detecting clubs and associations provide resources that help enthusiasts keep up to date with the legal aspects of their hobby. Joining such groups can provide valuable insights and updates on legislative changes, ensuring that your treasure hunting is both enjoyable and compliant with all applicable laws.

By thoroughly understanding and respecting local laws and regulations, metal detector users ensure that their hobby remains a sustainable and respected activity that can be enjoyed by future generations of enthusiasts. This legal compliance not only protects archaeological and historical sites but also enhances the reputation of the metal detecting community as a whole.

Respecting Private and Public Properties

Respecting private and public properties is a fundamental aspect of responsible metal detecting, encompassing both legal and ethical considerations. This respect not only ensures that metal detecting enthusiasts can continue to enjoy their hobby, but also preserves the integrity and privacy of the areas explored.

Firstly, the legal aspects of metal detecting on various properties can be complex and often vary by location. In the United States, laws regarding metal detecting are primarily determined at the state and local levels, with additional regulations sometimes imposed by specific agencies or municipalities. Before venturing out, it is essential for metal detectorists to thoroughly research and comply with all applicable laws. On public lands, such as state parks and beaches, permits may be required, and certain areas might be off-limits to preserve historical or ecological sites. Ignoring these laws can result in hefty fines, confiscation of equipment, and even criminal charges.

Private property, on the other hand, is governed by a different set of rules that require explicit permission from the property owner before detecting. Gaining this permission involves more than just avoiding legal repercussions; it establishes a relationship based on trust and mutual respect. Detectorists should ensure they are clear on any conditions set by the property owner and adhere strictly to those terms. Common stipulations might include restrictions on digging methods, specific areas where detecting is allowed, and agreements on the sharing or handling of any finds.

Ethically, respecting property goes beyond mere legal compliance. It involves practicing good stewardship of the land. This means following the principles of "Leave No Trace," which includes filling in all holes, disposing of any trash found (not just your own), and causing minimal disturbance to the natural setting. These practices help maintain the aesthetic and ecological integrity of the environment.

Additionally, the ethical responsibility extends to the handling of any items found. If a metal detectorist discovers items that could have historical or cultural significance, it is important to report such finds to local authorities or historical societies when required by law. In some cases, even if not legally mandatory, sharing information about significant finds with the community can contribute to the local historical knowledge base.

By maintaining a high standard of respect for both private and public properties, metal detectorists not only safeguard their ability to practice their hobby but also contribute positively to their communities and help ensure that historical and natural sites remain intact for future generations. This responsible approach helps to keep the hobby sustainable, enjoyable, and respectful of all parties involved.

Promoting Responsible Metal Detecting

Promoting responsible metal detecting involves a keen understanding of both legal and ethical considerations. These rules and principles ensure that hobbyists not only respect the law but also preserve historical and environmental integrity.

Firstly, understanding and adhering to local laws is paramount. In many places, metal detecting requires specific permissions, especially when conducted on public lands, parks, or historic sites. It is essential for enthusiasts to inquire with local authorities or land management offices before beginning their search, as some areas may be completely off-limits to metal detecting due to their historical or ecological significance. For example, detecting in national parks without explicit permission is illegal in the United States, and similar restrictions apply to state parks and historical sites. Additionally, some beaches may require a permit or restrict detecting to certain areas or times of the year.

The next aspect to consider is the ethical practice of metal detecting. This begins with respecting private property by always obtaining permission from landowners before detecting on their land. Trespassing not only tarnishes the image of the metal detecting community but can also lead to heavy fines and legal action.

When it comes to handling finds, responsible detectorists should follow the adage, "Take only memories, leave only footprints." This means carefully excavating items and restoring the site to its original condition as much as possible. This practice helps maintain the natural and aesthetic value of the environment. For items of potential historical value, it's important to report such finds to local authorities or historical societies, as these could contribute valuable information to the area's heritage.

Responsible metal detecting also involves promoting and practicing the reporting of finds when appropriate. In some jurisdictions, there may be legal requirements to report items that are over a certain age or of particular historical significance. By doing so, detectorists can support archaeological and historical research and preservation.

Another key component of ethical metal detecting is the consideration of wildlife and natural habitats. Detectorists should ensure that their activities do not disturb local wildlife or damage ecological areas, such as nesting grounds or native plant populations. This responsibility includes being mindful of the seasons during which wildlife is most vulnerable, such as breeding seasons.

Furthermore, responsible metal detecting advocates for the education and mentoring of new enthusiasts about best practices and legal obligations. This can be achieved through clubs, forums, and community groups, which often organize clean-up events or collaborate with local historical societies to help preserve and even enhance historical sites.

In conclusion, promoting responsible metal detecting is about much more than following the law. It's about fostering a community that values and actively contributes to the preservation of cultural heritage and the environment. By respecting these principles, metal detecting can continue to be a rewarding pastime that harmoniously coexists with cultural preservation and environmental conservation.

Chapter: 7 Advanced Techniques and Tips

Enhancing Detection Depth and Accuracy

Enhancing the depth and accuracy of detection when using a metal detector involves several advanced techniques and practical tips that can greatly improve the quality and success of your treasure hunts. Mastery of these skills not only increases the likelihood of discovering more valuable items buried deeper in the ground but also enhances the overall efficiency of searches.

One fundamental way to enhance detection depth is by choosing the right coil. Larger search coils generally increase the depth at which a detector can find larger objects, although they may reduce sensitivity to smaller items. Conversely, smaller coils are better at detecting smaller items and offer improved target separation in trashy areas. Swapping coils based on the specific conditions of your search area can be a strategic move.

Ground balancing is another critical factor for improving both depth and accuracy. This process involves adjusting the metal detector so it ignores the effects of naturally occurring minerals in the soil. Most high-end detectors offer automatic, manual, or tracking ground balance options. For highly mineralized soils, manually calibrating your detector ensures that it is not overly responsive to mineralization, which can mask the signals of deeper targets.

Fine-tuning sensitivity settings is also vital. While higher sensitivity increases the detector's ability to pick up signals from deeper objects, setting it too high can lead to false signals and chatter in mineral-rich soils. The key is to find the highest setting that allows stable operation without significant interference from ground mineralization or other environmental factors.

Discrimination settings can also impact depth and accuracy. Discrimination helps you avoid undesirable targets such as iron and pull tabs. However, setting the discrimination level too high might cause you to miss valuable targets that lie close to unwanted metals. Learning how to adjust these settings appropriately for different areas can greatly increase the effectiveness of your searches.

Using the all-metal mode can sometimes lead to deeper detections. This mode is more sensitive to all types of metal objects and provides the greatest depth of detection. Although it requires more patience due to the high volume of potential false signals, using all-metal mode in areas where valuable finds are more likely can be beneficial.

Pinpointing techniques also play a crucial role in enhancing accuracy. Once a target is roughly located, using a detector's pinpoint mode—if equipped—or a handheld pinpointer can help precisely locate the item, reducing unnecessary digging and disturbance to the surrounding area, thus preserving the condition of the find.

Regular practice and field testing different settings in varied environments is essential for mastering these advanced techniques. Every outing with your metal detector is an opportunity to refine your skills, learn how different settings behave under different conditions, and understand how subtle changes can affect the outcome of your searches.

By implementing these advanced techniques, metal detecting enthusiasts can not only increase their chances of uncovering hidden treasures but also make their experiences more rewarding and efficient. With patience, practice, and continual learning, you can significantly enhance both the depth and accuracy of your metal detector.

Using Metal Detectors in Different Environments

Using a metal detector effectively requires adapting techniques and settings to suit different environments, each presenting unique challenges and opportunities. From the mineralized wet sands of a beach to the trash-laden grounds of urban parks, metal detecting strategies can greatly influence the quantity and quality of finds.

On the Beach: Beaches are popular spots for metal detecting due to the high likelihood of finding lost jewelry and coins. When detecting on dry sand, standard settings on a Very Low Frequency (VLF) detector usually suffice. However, the wet sand and saltwater can cause false signals due to their mineral content. Using a Pulse Induction (PI) detector or a VLF detector with adjustable ground balance and sensitivity settings tailored for mineralized conditions can mitigate this. Sweeping the coil in a slow, methodical manner and overlapping each sweep ensures no area is missed. Early mornings or late evenings are ideal times as beaches are less crowded, and recent drops from the previous day are yet to be picked up.

In the Water: Water hunting can be rewarding but also challenging. Waterproof metal detectors are a must. PI detectors often excel in this environment due to their ability to ignore saltwater's conductivity. When searching in shallow water, ensure the coil is submerged and maintain a steady, slow swing of the detector. In deeper water, using a snorkel or diving gear along with an underwater detector will allow access to items lost offshore. Tides and currents can shift sand and uncover deeper items, so understanding tidal movements can help plan the most effective times to search.

Urban Areas: Urban environments are littered with metal trash which can make detecting a challenge. Using a detector with high discrimination settings helps filter out undesirable targets. A smaller coil can navigate around trash and focus on potential treasure spots like old walking paths, parks, and near historic buildings where old coins and relics may be found. Urban detecting often requires permissions from property owners or city authorities, so ensuring all legal aspects are covered is crucial before starting.

Rural and Wilderness Areas: These areas can be a treasure trove for relic hunters. Old farms, abandoned homesteads, and areas that once saw historical activity are prime spots. A metal detector with a larger coil can cover vast areas more effectively, and settings should be adjusted for lower discrimination to pick up a broader range of target signals. Rural detecting often involves dealing with more varied terrain, so using a detector with manual ground balance adjusts to different soil types, from high mineral soil to organic-rich loam.

Advanced Techniques and Tips: Regardless of the environment, mastering the metal detector's threshold sound—the faint background hum—can help distinguish between deeper, fainter signals and ground noise. Regularly practicing with a test garden, where various objects are buried at known depths, can train the ear to recognize different types of signals and improve digging accuracy. Keeping a log of finds and detector settings used at different sites can provide insights into what works best in various conditions. Moreover, always be prepared with the right gear for the environment—waterproof boots for wet areas, a hat and sunscreen for beach hunts, and sturdy gloves for digging in urban and rural areas.

Adapting to each environment not only increases the likelihood of successful finds but also makes metal detecting a more enjoyable and rewarding hobby. Each location offers a unique history and a different set of potential treasures just waiting to be uncovered.

Participating in Competitions and Events

Participating in metal detecting competitions and events is a thrilling way to enhance your skills, meet other enthusiasts, and possibly win some prizes. These events range from friendly local gatherings to large, organized competitions where specific rules and challenges test the abilities of participants.

When you decide to join a metal detecting competition, it's crucial to first understand the type of event you're entering. Competitions may vary widely in their structure. Some might focus on who can find the most items within a set period, while others might reward the discovery of the most valuable or oldest items. There are also themed hunts, where specific types of items are the targets, such as historical relics or natural gold.

Preparation for these events involves more than just knowing how to use your metal detector. It's essential to prepare mentally and physically. Most competitions take place outdoors, often requiring participants to be on their feet for extended periods across varied terrains. Physical stamina and the ability to navigate these environments comfortably can make a significant difference in your performance.

Knowing your equipment is paramount. Competitors should spend ample time with their metal detectors before the event to understand every feature and setting. Familiarity with adjusting the detector's sensitivity, discrimination, and ground balance settings can lead to quicker and more accurate finds during the competition. Advanced users might also benefit from using a pinpointer—a handheld device that helps locate the exact position of an object in the ground, speeding up recovery time.

Researching the event location can provide a competitive edge. If historical research or previous finds are available, they can give clues about what might be found and at what depths. For beach events, understanding tides and how they affect sand movement can influence where to search. For historical land sites, knowing past usage of the land can guide what settings to use on your detector and where to focus your search efforts.

Respecting the rules of the competition is crucial. These might include restrictions on where you can search, what tools you can use for digging, and how to handle finds. Some events require that holes be refilled and the site left as it was found, promoting responsible metal detecting practices.

Networking with other participants offers learning opportunities and can enhance the overall experience. More experienced competitors often share insights and techniques that they have perfected over years of detecting. These interactions can also lead to invitations to other hunts or tips on other productive sites.

Finally, understanding the etiquette and sportsmanship expected at these events ensures that everyone has a fair and enjoyable experience. This includes respecting other competitors' space, not encroaching on their designated area, and being honest about your finds.

By participating in metal detecting competitions and events, enthusiasts not only put their skills to the test but also contribute to the community's knowledge. Each competition can be a learning experience, offering insights into metal detecting techniques, historical knowledge, and the thrill of the hunt.

Chapter: 8 Applications of Metal Detecting

Treasure Hunting

Treasure hunting with a metal detector is a thrilling hobby that combines outdoor adventure with the chance to discover hidden treasures and historical artifacts. It's an activity that appeals to enthusiasts of all ages, offering both a relaxing pastime and an exciting quest for valuable finds. By understanding the different applications of metal detecting and mastering the use of this equipment, treasure hunters can significantly increase their chances of successful finds.

One of the primary appeals of metal detecting is the possibility of uncovering lost or forgotten items that have historical, sentimental, or monetary value. This could range from ancient coins and military relics to jewelry and old toys. Each find provides a glimpse into the past, potentially offering new insights into local history and the daily lives of people who lived in different eras. For many treasure hunters, each item unearthed is a story waiting to be told, making metal detecting a particularly fulfilling hobby for history buffs.

To embark on treasure hunting with a metal detector, it's crucial to understand the capabilities and limitations of the device. Metal detectors can vary widely in their functionalities, including depth of detection, sensitivity to different metals, and the ability to discriminate against undesirable objects. High-frequency detectors, for instance, are better suited for finding small objects like gold nuggets or fine jewelry, while lower frequency detectors can cover more ground and are ideal for locating larger items like cannonballs or large metal containers.

The environment plays a significant role in metal detecting. Beaches, old battlefields, abandoned homesteads, and even backyards can be rich hunting grounds. Each of these locations requires different techniques and settings on the metal detector. For example, wet sand on beaches or highly mineralized soil conditions can affect the detector's performance, necessitating adjustments to the ground balance settings to reduce noise and improve target identification.

Proper research can greatly enhance the effectiveness of a treasure hunt. Historical research to identify areas that were once populated or used for specific activities can lead to more targeted and productive searches. Old maps, historical documents, and records can point to forgotten settlements, old fairgrounds, or areas that once hosted significant public events, all of which are likely spots for finding relics and coins.

The legal aspect of treasure hunting must also be carefully considered. Regulations regarding metal detecting vary by location, with some areas requiring permits or completely restricting metal detecting activities. Additionally, ethical practices such as obtaining permission to search on private land, respecting property, and carefully restoring any disturbed areas are fundamental to responsible treasure hunting.

Ultimately, the success in treasure hunting with a metal detector hinges on a blend of skillful use of the equipment, knowledge of historical contexts, and an understanding of legal and ethical considerations. For those willing to invest the time in learning and practice, metal detecting can be a rewarding hobby that not only brings the thrill of discovery but also offers a unique way to connect with history and nature.

Historical Research

Metal detecting can be a valuable tool for historical research, offering a unique method to uncover and preserve artifacts that provide insight into the past. When used effectively, metal detectors help historians, archaeologists, and enthusiasts discover objects that tell stories of bygone eras, from ancient civilizations to more recent historical events.

The application of metal detecting in historical research often starts with careful planning and research. Users need to study historical records, maps, and documents to identify promising locations where settlements, battles, or everyday activities may have occurred. This preparatory work is crucial because it directs the search to areas with high potential for significant finds.

Once a site is chosen, metal detecting can begin. The technology is particularly adept at locating metal objects such as coins, jewelry, tools, weapons, and other personal or utilitarian items. Each find has the potential to offer insights into the material culture of a specific time period. For example, coins can help identify the era of activity at a site and can even pinpoint economic conditions through mint dates and circulation patterns.

In historical research, the ability to differentiate between types of metal and the depth at which objects are found is invaluable. This is where knowledge of metal detector settings and capabilities becomes important. Settings such as discrimination, sensitivity, and ground balance must be adjusted according to the soil composition and expected artifact types. Such adjustments help minimize the recovery of irrelevant items and enhance the chances of unearthing historically significant artifacts.

Proper excavation techniques also play a critical role in the process. When a potential artifact is detected, careful excavation is necessary to ensure that the item and its surrounding context are preserved. This approach allows researchers to maintain the integrity of the site and the artifacts, which is essential for accurate analysis and documentation.

The finds must then be cleaned, catalogued, and studied. Researchers often collaborate with specialists in various fields to analyze the objects, using techniques like metallography or dating methods to learn more about their composition and history. The results of these studies can contribute valuable information to our understanding of historical events, trade relations, and daily life in different periods.

Moreover, metal detecting can sometimes lead to discoveries that challenge existing historical narratives or fill in gaps in our knowledge. For example, finding artifacts from a previously unknown military encampment can alter our understanding of historical events and lead to new interpretations of documented battles or movements.

However, it's crucial for metal detector users engaged in historical research to operate within the framework of local laws and ethical guidelines. Many regions have strict regulations governing the use of metal detectors, especially on land with archaeological significance. Responsible metal detectorists must obtain necessary permissions, report their finds where required, and often work in collaboration with professional archaeologists.

In summary, metal detecting can significantly enrich historical research by providing direct access to physical artifacts from the past. When conducted responsibly and methodically, it bridges the gap between theoretical historical studies and tangible evidence, bringing history to life in a way that few other research methods can.

Geological Exploration

Geological exploration is a fascinating application of metal detecting that goes beyond the usual hunt for coins, jewelry, and historical relics. In this context, metal detectors become powerful tools for identifying and locating mineral deposits and geological features. Understanding how to use a metal detector effectively in geological exploration involves grasping both the scientific principles behind the device and the practical techniques for fieldwork.

Metal detectors used in geological exploration are typically designed to be sensitive to the specific conductivity and magnetic properties of target metals such as gold, silver, or platinum. These devices are often used in areas where high mineralization challenges other types of electronic devices. A metal detector used for geological purposes must be capable of adjusting to highly mineralized soils, distinguishing between different types of metal, and operating in varied terrains.

The process begins with selecting a suitable metal detector. Most geological detectors are of the Very Low Frequency (VLF) or Pulse Induction (PI) type, with PI models being particularly beneficial in gold-rich areas where their depth and sensitivity to small nuggets are invaluable. Advanced features like ground balance are crucial; this allows the detector to compensate for minerals in the soil, thereby reducing noise and increasing the likelihood of detecting metal objects.

Once in the field, operators must be proficient in using their detectors. This includes knowing how to calibrate the device according to the specific ground conditions, which can vary even within a small area. Skilled users will often test their detectors on samples of local rocks and minerals to fine-tune the settings for optimal sensitivity and discrimination. Effective sweeping techniques, such as maintaining a consistent coil height and using slow, overlapping sweeps, enhance the chances of uncovering hidden metallic objects.

Interpreting the signals from the metal detector is another critical skill. Geological explorers need to discern between different tones and identify patterns that indicate the presence of a targeted metal. For instance, the distinct sound or digital reading produced when a detector sweeps over a gold nugget differs significantly from that of iron or lead, and recognizing these differences can save considerable time and effort.

For geologists and prospectors, metal detectors are also used in conjunction with other tools and methods. Mapping geographic features, sampling soils, and analyzing geological formations are all part of a broader strategy to understand the area's potential for mineralization. Metal detecting provides a means of quickly surveying an area, allowing explorers to focus their sampling efforts more effectively.

In summary, the application of metal detecting in geological exploration requires a deep understanding of both the technology and the terrain. It is a skill that combines the scientific method with practical experience, allowing geologists and prospectors to locate new resources and expand our understanding of the Earth's mineral wealth. This makes mastering a metal detector not just a hobby but a professional skill for those involved in the exploration of natural resources.

Recovery and Rescue Missions

Metal detectors play a critical role in recovery and rescue missions, providing an essential tool for locating metallic objects that are vital in various search operations. These devices are particularly useful in scenarios ranging from finding evidence in criminal investigations to searching for missing artifacts or historical relics. Moreover, in the aftermath of disasters, metal detectors can be indispensable for locating buried survivors or identifying hazardous materials.

In recovery missions, metal detectors are commonly used by law enforcement and forensic teams. For example, after a crime, investigators may need to recover discarded weapons, shell casings, or other metallic evidential items from crime scenes. The sensitivity of metal detectors makes them ideal for sweeping large areas and pinpointing the location of items that are crucial for forensic analysis. By adjusting the settings to minimize interference from other metals, users can focus specifically on the type of object they are tasked to find, thus enhancing the efficiency of the search and the integrity of the evidence collected.

In the context of archaeological digs or historical explorations, metal detectors help in uncovering artifacts that tell stories of past civilizations. Archaeologists rely on the technology to locate objects without disturbing too much soil or causing damage to hidden treasures. Here, metal detectors must be used with great care and precision, often set to detect specific types of metals while ignoring others that may be commonly found in the soil but are not of historical interest.

Rescue missions, especially following natural disasters such as earthquakes, hurricanes, or avalanches, can also benefit from metal detection technology. In these scenarios, metal detectors help rescue teams identify locations where people may be trapped under debris. This is possible because metal detectors can locate personal belongings such as jewelry, phones, or even belt buckles, which might indicate the presence of a person. Advanced metal detectors used in these operations are designed to penetrate deeply buried layers of rubble and provide real-time data to rescuers, significantly speeding up the search process and increasing the chances of saving lives.

Additionally, metal detectors are employed in the cleanup and recovery efforts in areas affected by industrial accidents or wars. They are instrumental in locating and removing hazardous debris like unexploded ordnance in former battlefields or detecting metallic contaminants in environments that have suffered industrial spills. In these applications, the ability of metal detectors to function in highly mineralized soils and under challenging environmental conditions is crucial.

The effective use of metal detectors in recovery and rescue missions depends not only on the device itself but also on the operator's expertise. Mastery over the device's settings—such as sensitivity, discrimination, and ground balance—is essential to adapt to different scenarios and targets. Operators must also be trained in search patterns and techniques to maximize coverage and ensure no crucial evidence or object is overlooked. As metal detectors become more sophisticated, their impact on recovery and rescue missions continues to grow, making them indispensable tools in both emergency response and systematic exploratory or investigative operations.

Chapter: 9 Case Studies and Success Stories

Notable Finds

Metal detecting is a hobby that occasionally turns up much more than mere coins or discarded bottle tops. Across the globe, enthusiasts have unearthed objects of significant historical, cultural, or financial value. These finds not only enrich our understanding of history but also illustrate the potential rewards of mastering the use of a metal detector.

One of the most inspiring stories is that of Terry Herbert, who in 2009, using a relatively inexpensive metal detector, discovered the Staffordshire Hoard in England. This treasure, valued at over 3 million pounds, is the largest hoard of Anglo-Saxon gold and silver metalwork ever found. Terry's discovery, which included over 3,500 items, some of which are believed to date back to the 7th century, highlights not just the potential financial gain from this hobby but also the immense historical significance such finds can represent.

In the United States, a metal detecting enthusiast named Mike DeMar, while vacationing in Florida, found a gold pendant that turned out to be a piece of Spanish history. The pendant, adorned with an image of the Virgin Mary, was part of a 1715 shipwreck's treasure off the Florida coast. This remarkable find was appraised at approximately $500,000. Mike's success underscores the importance of location in metal detecting, encouraging enthusiasts to consider the historical context of their search areas.

Another noteworthy find was made by a novice detectorist in Scotland, who unearthed the Galloway Hoard. This Viking-era treasure trove included over 100 gold and silver objects dating back more than 1,000 years. The discovery not only provided a significant personal triumph but also offered invaluable insights into Viking presence in Scotland. The find demonstrates the potential for significant historical discoveries that can contribute to our understanding of early medieval history.

These stories share common themes: the importance of patience, persistence, and a deep understanding of one's metal detector. Each of these finders had invested time in learning how to adjust their detector's settings according to the ground conditions and how to interpret the machine's signals accurately. Moreover, they researched and chose their hunting grounds wisely, often selecting areas known for historical activity or previous finds.

These case studies not only serve as motivational tales for aspiring treasure hunters but also highlight the crucial role of using metal detectors effectively. They show that while luck certainly plays a part, knowledge of the technology and the history of potential search sites are equally important. Whether the aim is to find precious metals, historical artifacts, or simply to enjoy the thrill of the hunt, these stories illustrate the rich rewards that metal detecting can offer.

Interviews with Experienced Metal Detecting Enthusiasts

Interviews with experienced metal detecting enthusiasts reveal a fascinating mosaic of success stories, practical advice, and insights into the treasure hunting community. These case studies provide not only motivation but also invaluable lessons on how to navigate the hobby of metal detecting, often emphasizing the importance of persistence, knowledge, and the right equipment.

One notable story comes from Carol, a seasoned detectorist from California, who discovered a rare 19th-century coin in a forgotten picnic area near the Sierra Nevada. Her success is attributed to her methodical approach to researching historical records and old maps to pinpoint potential hotspots. Carol stresses the importance of understanding the history of a location to maximize the chances of finding something valuable. She advises new enthusiasts to spend as much time researching as they do detecting, emphasizing that knowledge of historical activity can lead to successful finds.

Another enthusiast, Jim from Florida, shares a different perspective focused on beach and underwater detecting. Specializing in searching along the hurricane-prone coasts, Jim uses a pulse induction metal detector designed to perform in highly mineralized soils like wet salt sand and submerged environments. His most significant find, a Spanish doubloon from a shipwreck, was located after a storm shifted the sandy seabed. Jim highlights the necessity of understanding your metal detector's settings and capabilities, particularly in challenging conditions where typical VLF detectors might falter. He suggests investing in equipment that can handle the specific environments where you plan to search and taking the time to practice and master its use under various conditions.

Ellen, a detectorist from Vermont, focuses on the role of community and networking in metal detecting. Her story of uncovering a cache of colonial-era artifacts was made possible through connections with other hobbyists who shared tips and locations. Ellen emphasizes attending club meetings, participating in forums, and engaging with the community both online and in person. According to her, these interactions not only enhance one's detecting skills through shared knowledge but also often lead to partnerships and opportunities to explore new locations.

These enthusiasts also discuss the unexpected benefits of metal detecting, such as improving physical fitness, enjoying the outdoors, and the therapeutic aspects of spending hours in nature focused on a task. They share tales of excitement, disappointment, and sometimes, spectacular recoveries, which keep the passion alive.

Beyond individual successes, these stories often touch on ethical practices in metal detecting. This includes obtaining proper permissions to search on private property, respecting historical sites, and responsibly handling any items found, particularly those of historical significance. They advocate for following the code of ethics promoted by metal detecting associations, such as filling in holes and properly disposing of trash uncovered during searches.

Interviewing experienced metal detecting enthusiasts offers a rich vein of knowledge and experience. For anyone serious about using a metal detector, these insights and stories are not just inspirational—they're practical guides that demonstrate the blend of patience, skill, and respect for history and community that defines successful detectorists.

Chapter: 10 Resources and Further Reading

Recommended Books and Guides

For those eager to delve deeper into the world of metal detecting, a wide range of literature is available that covers everything from the basics of operation to the subtleties of advanced techniques. These books and guides not only enhance practical knowledge but also enrich the treasure hunting experience with historical insights and expert tips.

One highly recommended resource is "The Metal Detecting Bible" by Brandon Neice. This book offers a comprehensive look at various aspects of metal detecting, from choosing the right equipment to understanding where to search and how to clean and preserve finds. Its practical approach makes it invaluable for both beginners and seasoned enthusiasts.

Another essential guide is "Metal Detecting for the Beginner" by Vince Migliore. It is particularly suited to those just starting out, providing straightforward advice on the technical aspects of detectors and detailed guidance on how to begin one's journey in metal detecting. This book also tackles the common mistakes novices make, offering solutions that save time and improve efficiency.

For those interested in the historical and archaeological perspectives of metal detecting, "The Handbook of Metal Detecting" by David Villanueva outlines not only the practical uses of metal detectors but also discusses how enthusiasts can contribute to historical research. The book encourages a responsible approach to treasure hunting, emphasizing the importance of ethical practices and cooperation with archaeological communities.

"Advanced Metal Detecting: Tips, Tricks and Techniques" by Daniel Bernzweig dives deeper into the technical specifics and advanced strategies that can help proficient detectorists enhance their skill set. This guide covers advanced ground balancing methods, techniques for dealing with highly mineralized soils, and tips for underwater detecting, making it a valuable resource for those looking to push the boundaries of their hobby.

For those who prefer a more narrative approach, "Lost Treasure Trails" by Thomas Penfield combines engaging stories with practical advice, recounting thrilling tales of treasure hunts while providing tips that can be applied by readers in their own searches. This book captures the excitement and allure of treasure hunting, motivating readers to explore and discover.

In addition to these books, numerous online forums and websites offer a wealth of information, ongoing discussions, and updates in the field of metal detecting. Sites like TreasureNet and The Friendly Metal Detecting Forum are communities where enthusiasts can share finds, seek advice, and discuss various aspects of metal detecting. These platforms also often feature downloadable guides, e-books, and articles that are updated regularly with the latest information and techniques.

Each of these resources provides valuable insights that can help enhance one's understanding of metal detecting, from handling the equipment and respecting historical artifacts to mastering the art of finding hidden treasures. Whether you are a beginner looking to get a solid start or an experienced detectorist aiming to refine your skills, these books and guides are essential tools in your metal detecting arsenal.

Useful Websites and Forums

For enthusiasts of metal detecting, whether novice or experienced, the internet offers a wealth of resources that can enhance their understanding of the hobby and improve their detecting skills. Useful websites and forums provide platforms where individuals can share experiences, learn from others, find answers to specific questions, and stay updated on the latest equipment and techniques.

Websites dedicated to metal detecting often serve as comprehensive hubs for all things related to the hobby. These sites may include tutorials, articles, and reviews of different metal detectors, which are invaluable for making informed purchasing decisions. Additionally, many of these websites offer video content, showcasing metal detecting adventures and providing visual and practical tips on how to use different detectors effectively. For example, websites like MetalDetector.com and SeriousDetecting.com not only sell detectors but also feature guides, how-to articles, and community updates that are crucial for both beginners and seasoned enthusiasts.

Forums play a crucial role in the metal detecting community by fostering interaction among members. These platforms allow users to post their finds, seek advice on detector usage, and discuss various aspects of their hobby, such as identifying unknown finds or best practices for securing permission to detect on private property. TreasureNet.com and DetectorProspector.com are two of the most popular forums where stories and information are exchanged daily. These forums are often segmented into categories based on detector brands, types of metal detecting (such as beach hunting or relic hunting), and locations, making it easier for users to navigate and engage with content that matches their specific interests.

Another valuable resource is the websites of national and regional metal detecting clubs. These sites often list upcoming events, club hunts, and meetings, which can be great opportunities for hands-on learning and networking. Clubs often host guest speakers who are experts in the field, offering insights into everything from historical research to the latest detecting technologies. Websites like AmericanMetalDetecting.com aggregate information about clubs across the United States, providing contact information and links to individual club websites.

In addition to these, online learning platforms and YouTube channels specifically dedicated to metal detecting can also be useful. Channels such as "Aquachigger" and "The Hoover Boys" not only entertain but also educate viewers by sharing their detecting trips and tips, from choosing the right gear to cleaning and preserving finds. These channels often review new equipment, demonstrating their use in the field, which can be extremely helpful for those looking to upgrade their gear or learn more about specific models.

The combination of educational content, community support, and up-to-date information found on these websites and forums makes them indispensable tools for anyone interested in metal detecting. They not only help users become more skilled with their equipment but also deepen their appreciation for the hobby by connecting them with a wider community of like-minded individuals. Whether you are looking for the thrill of discovery or the pleasure of historical research, these online resources can greatly enhance your metal detecting experience.

Training and Certification Programs

For those eager to master the art of metal detecting, diving into training and certification programs can significantly enhance your skills and understanding of this rewarding hobby. These programs are designed not only to teach the practical aspects of using metal detectors but also to imbue learners with a thorough knowledge of historical contexts, legal considerations, and ethical practices.

Training programs in metal detecting typically cover a wide range of topics, starting from the very basics of operating the different types of metal detectors, such as Very Low Frequency (VLF), Pulse Induction (PI), and Beat Frequency Oscillation (BFO). Participants learn how to adjust settings like sensitivity and discrimination, and how to interpret the signals their devices emit. This foundational knowledge is crucial for effective searching and can greatly increase the rate of significant finds.

More advanced training often includes modules on the geographical application of metal detecting. For instance, training might cover specific techniques for beach searching, where saltwater and wet sand pose unique challenges, or for relic hunting in historical sites, where understanding soil composition and historical timelines can lead to successful finds. These programs also typically incorporate hands-on field training, which not only puts theory into practice but also helps to refine the detectorist's technique under the guidance of experienced instructors.

Certification programs, while less common than general training courses, offer a formal acknowledgment of a detectorist's proficiency. These certifications can be particularly valuable for those looking to turn a hobby into a part-time or full-time profession, such as becoming a recovery specialist for hire, or for those engaging in historical preservation efforts. Certification can also lend credibility when seeking permission to search on private properties or when participating in archaeological digs.

Both training and certification programs usually emphasize the importance of ethical metal detecting. This includes training on how to seek the necessary permissions, how to report finds when required, and how to ensure minimal disturbance to the environment. Many programs also teach the legalities of metal detecting, which vary significantly from one region to another, covering topics such as trespass laws, national and state park regulations, and the specifics of the Archaeological Resources Protection Act.

Many community colleges, adult education centers, and even specialized metal detecting clubs offer these training programs. They can range from one-day workshops to more extensive courses spanning several weeks. For those unable to attend in-person training, there are also online courses and webinars available from various metal detecting societies and experienced detectorists who share their knowledge through digital platforms.

Additionally, further reading and resources are recommended for those pursuing a deeper understanding or specific aspects of metal detecting. Books on the history of metal detecting, guides to identifying historical artifacts, and manuals on the technical aspects of metal detectors are invaluable. Online forums and communities also provide ongoing support and a wealth of shared knowledge that is especially beneficial for both beginners and experienced enthusiasts alike.

Engaging with these training and certification programs not only enhances the technical skills of a metal detectorist but also ensures that their pursuits are conducted responsibly and sustainably, preserving the integrity of the sites explored and contributing positively to the community.

Conclusion

Concluding a guide on how to use a metal detector encapsulates not only the diverse aspects and techniques of using the device effectively but also reflects on the broader implications of engaging with this hobby. It's an activity that intertwines adventure, historical exploration, and often, a sense of community. Mastery of a metal detector doesn't happen overnight; it requires patience, practice, and persistence. With each outing, users deepen their understanding of their equipment and refine their techniques, which increases their chances of finding valuable or historically significant items.

In the process of learning to use a metal detector, enthusiasts come to appreciate the subtleties of their equipment—the way different soils affect the machine's responses, how weather and environmental conditions can alter its effectiveness, and how the settings need to be adjusted according to the specific location. They also gain an understanding of the rich history beneath their feet, often developing a greater appreciation for the preservation of cultural heritage.

The practical knowledge of operating the detector is just the beginning. The ethics of metal detecting play a crucial role in ensuring that enthusiasts respect the laws and guidelines, which are designed to protect archaeological and natural sites. This respect ensures that metal detecting remains a hobby that can be enjoyed by future generations and maintains the goodwill of the general public and governing bodies.

Furthermore, metal detecting can have a significant communal aspect. Clubs and online forums are not just about sharing tips and finds but also about fostering a community spirit that encourages responsible detecting. These communities often organize group hunts, participate in historical preservation efforts, and even assist law enforcement in recovery missions.

As technology advances, so too do the capabilities of metal detectors. Innovations in sensor and filtering technology allow detectors to be more selective in what they find and more accurate in their targeting. This progression makes the hobby continually exciting and rewarding for both new and seasoned enthusiasts.

In conclusion, learning to use a metal detector effectively opens up a world of exploration and discovery. It encourages historical education, fosters respect for the law and the environment, and builds community among those who share this passion. Whether the goal is to find treasures, explore history, or simply enjoy a day outdoors, metal detecting is a hobby that offers ongoing challenges and rewards. As users grow in their skills and knowledge, they contribute not only to their personal satisfaction but also to the larger community of enthusiasts and historians. This interplay of personal achievement and community engagement highlights the enriching nature of metal detecting, making it a fulfilling pursuit for people of all ages.

www.ingramcontent.com/pod-product-compliance
Lightning Source LLC
Chambersburg PA
CBHW050321230526
45471CB00005B/2284